I can cut it.

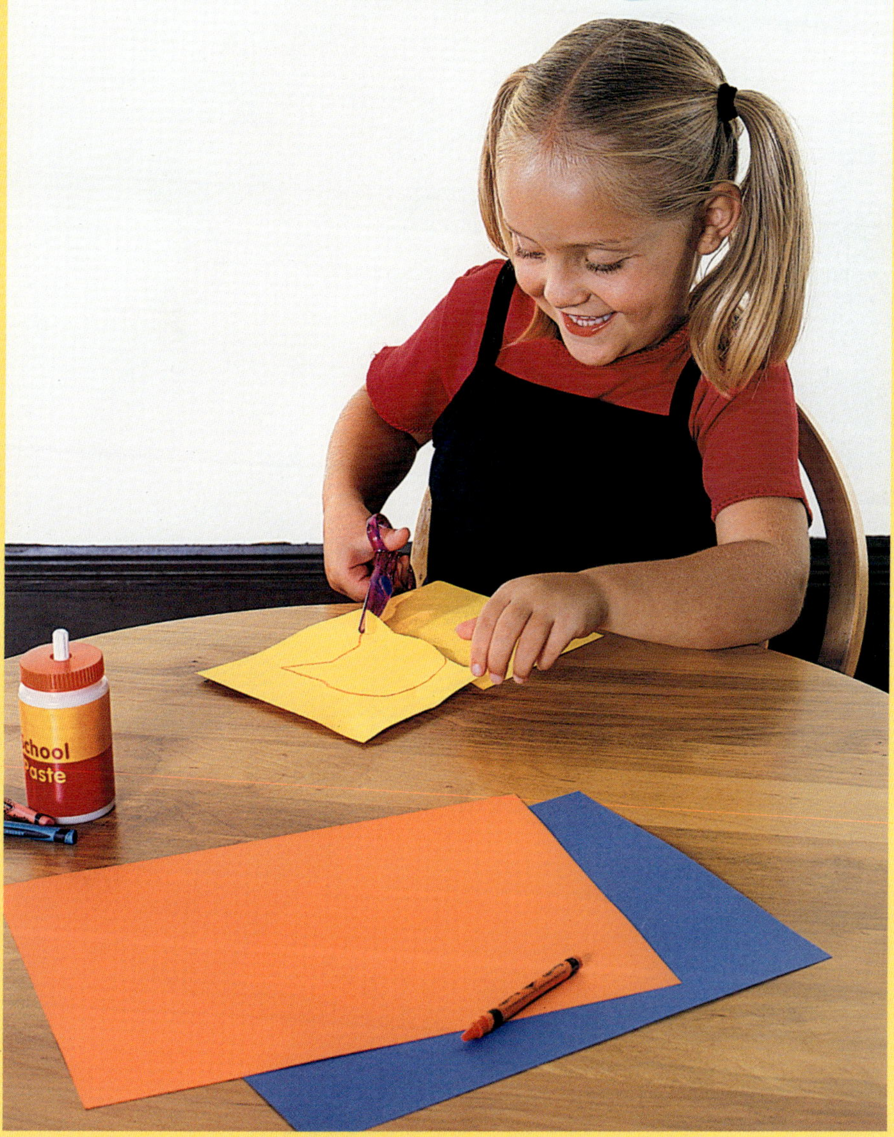

Is it a cat?

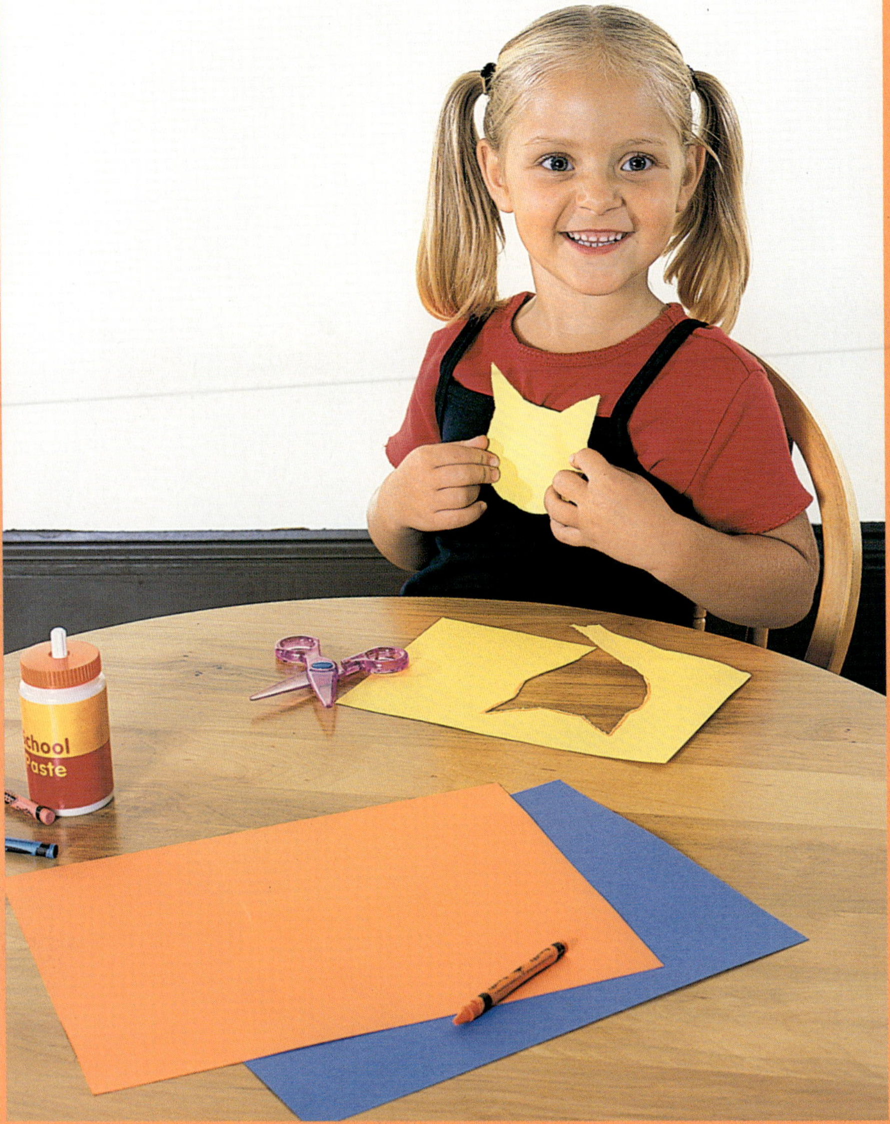

I can cut, cut, cut.

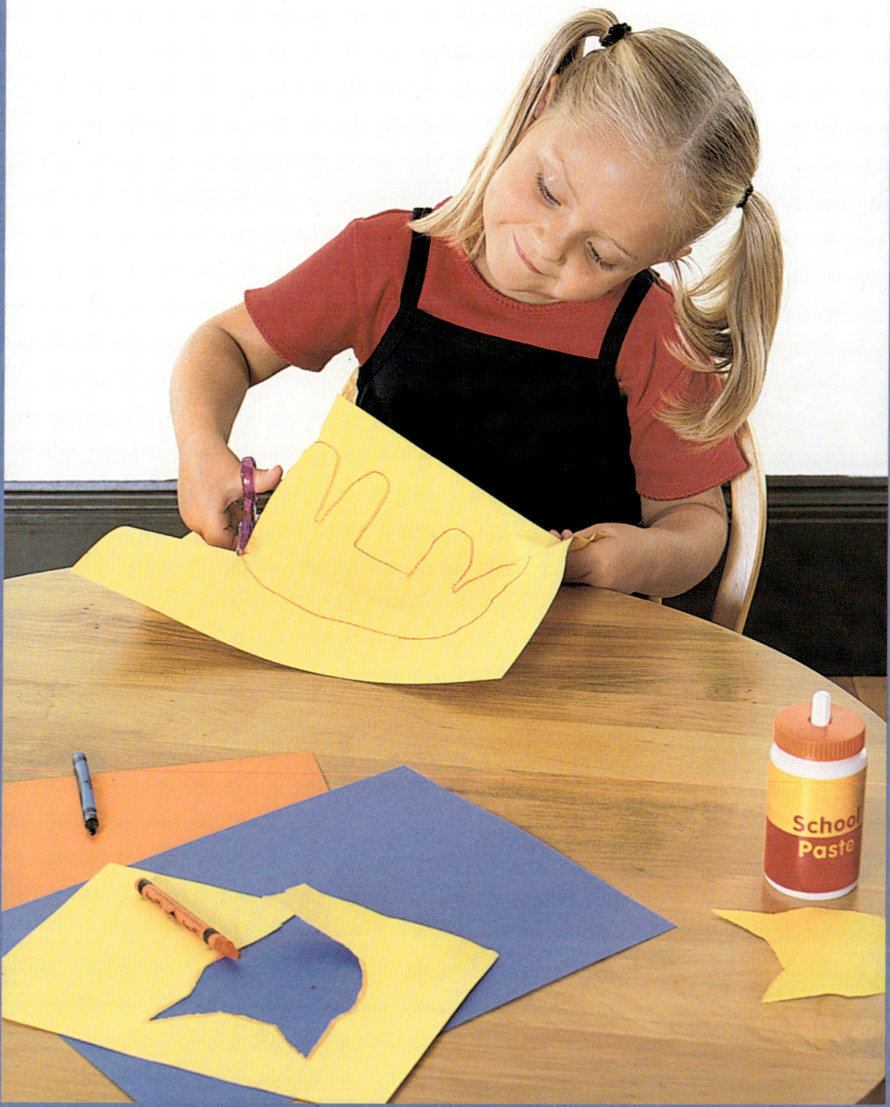

Is it a cat?

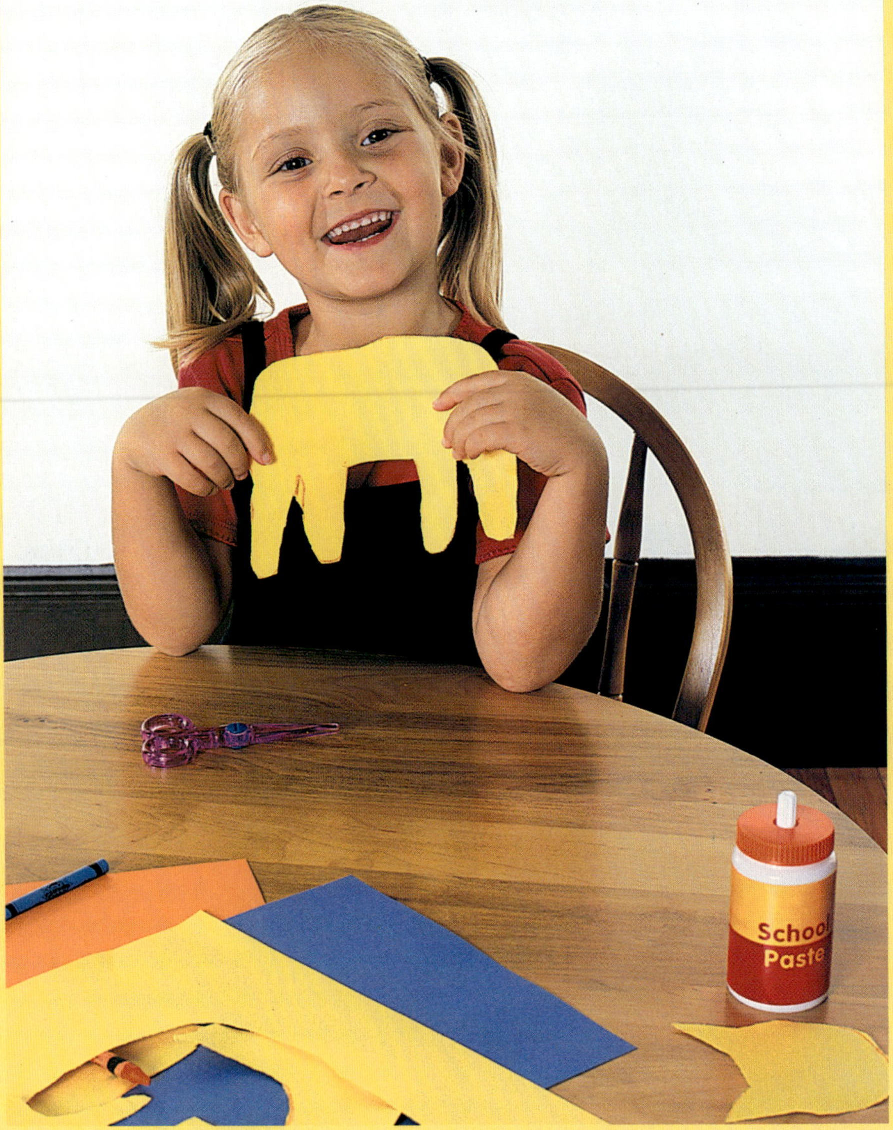

I can snip, snip, snip.

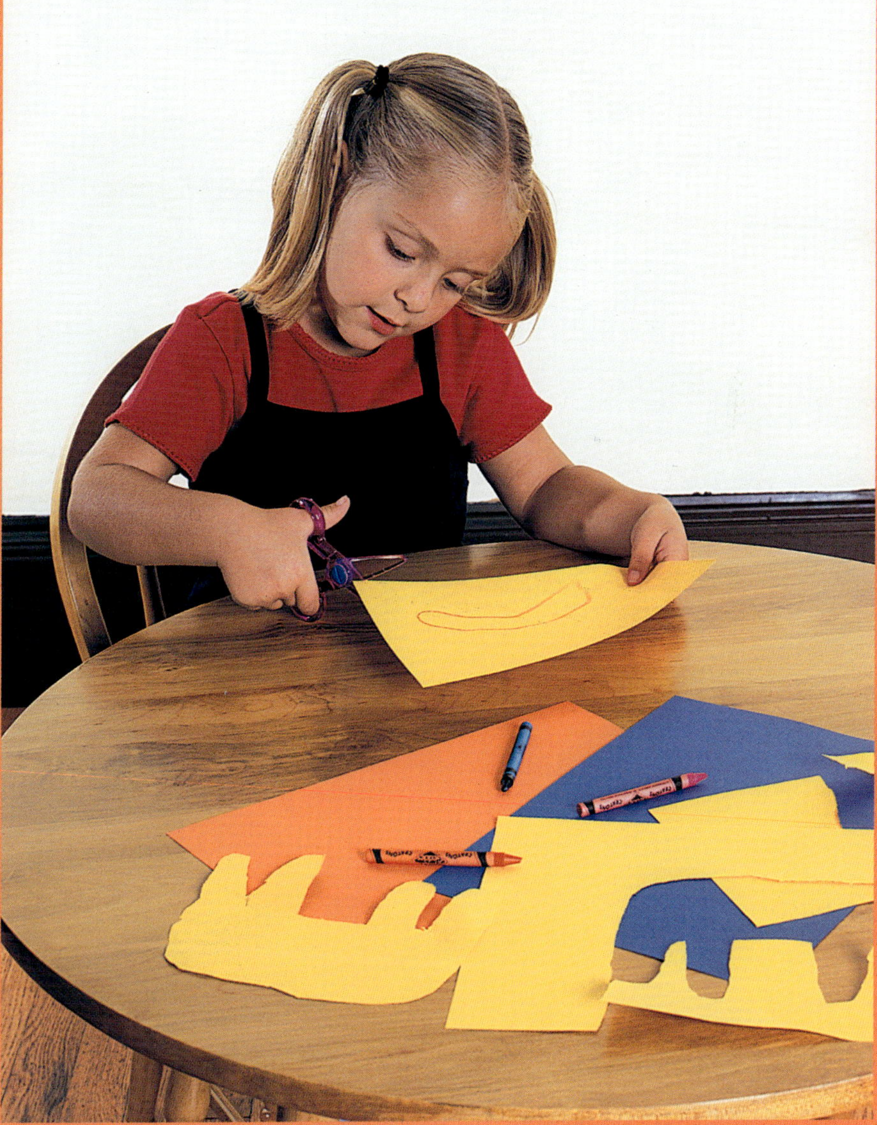

Is it a cat?

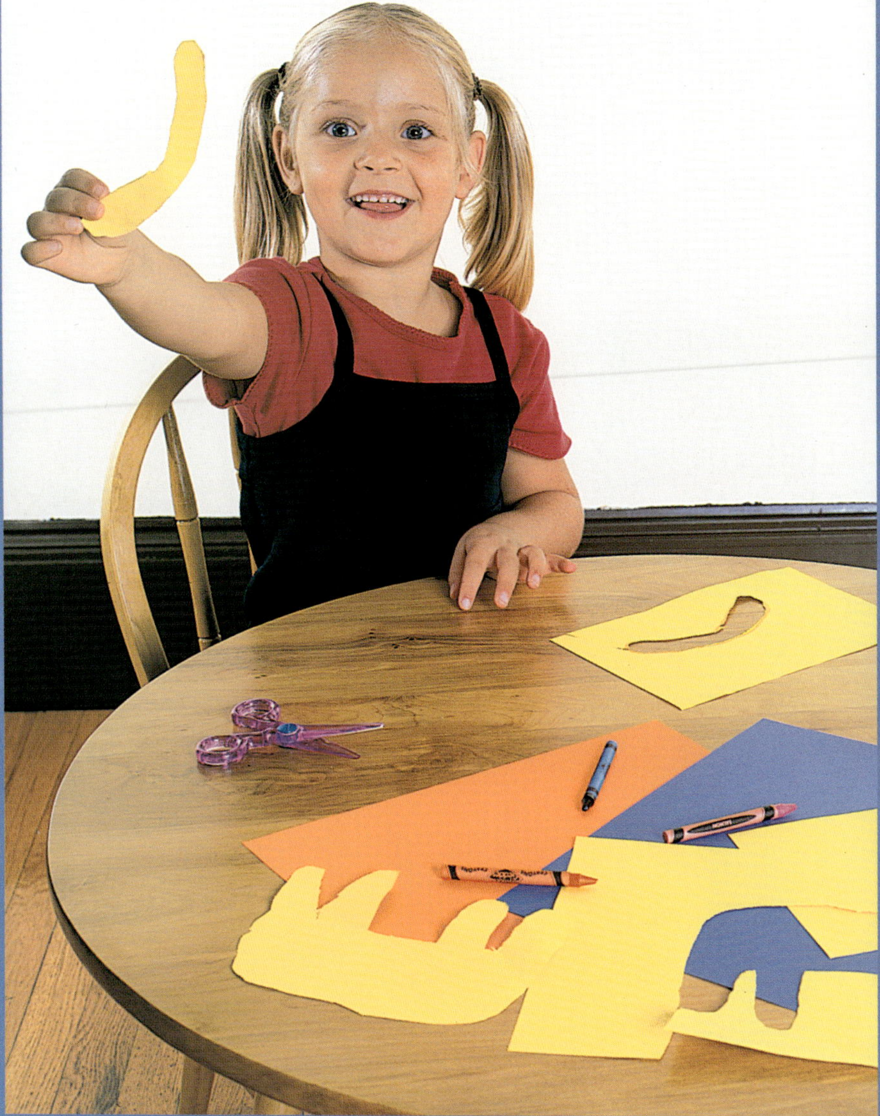

I fit it into a 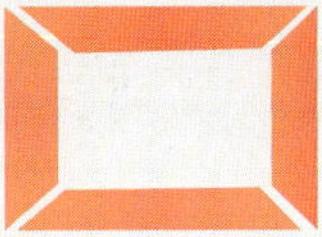.

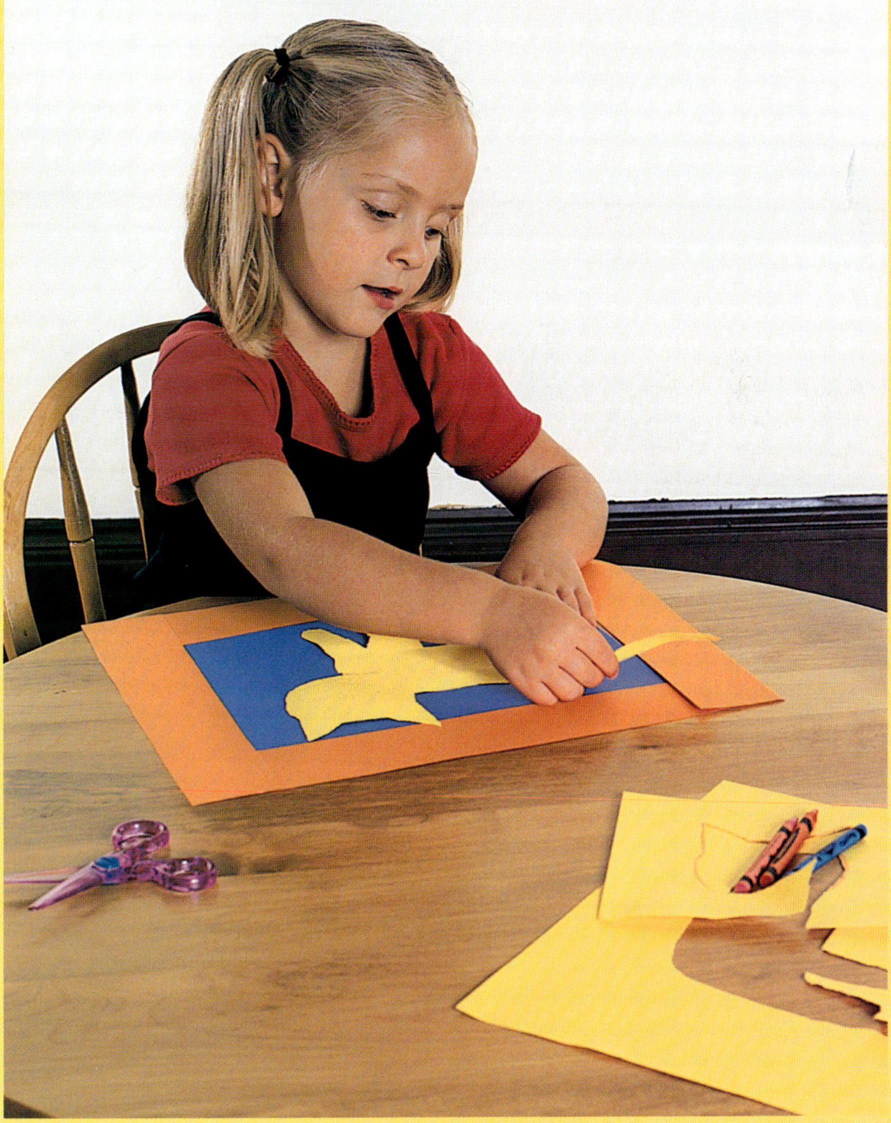

It is a cat!

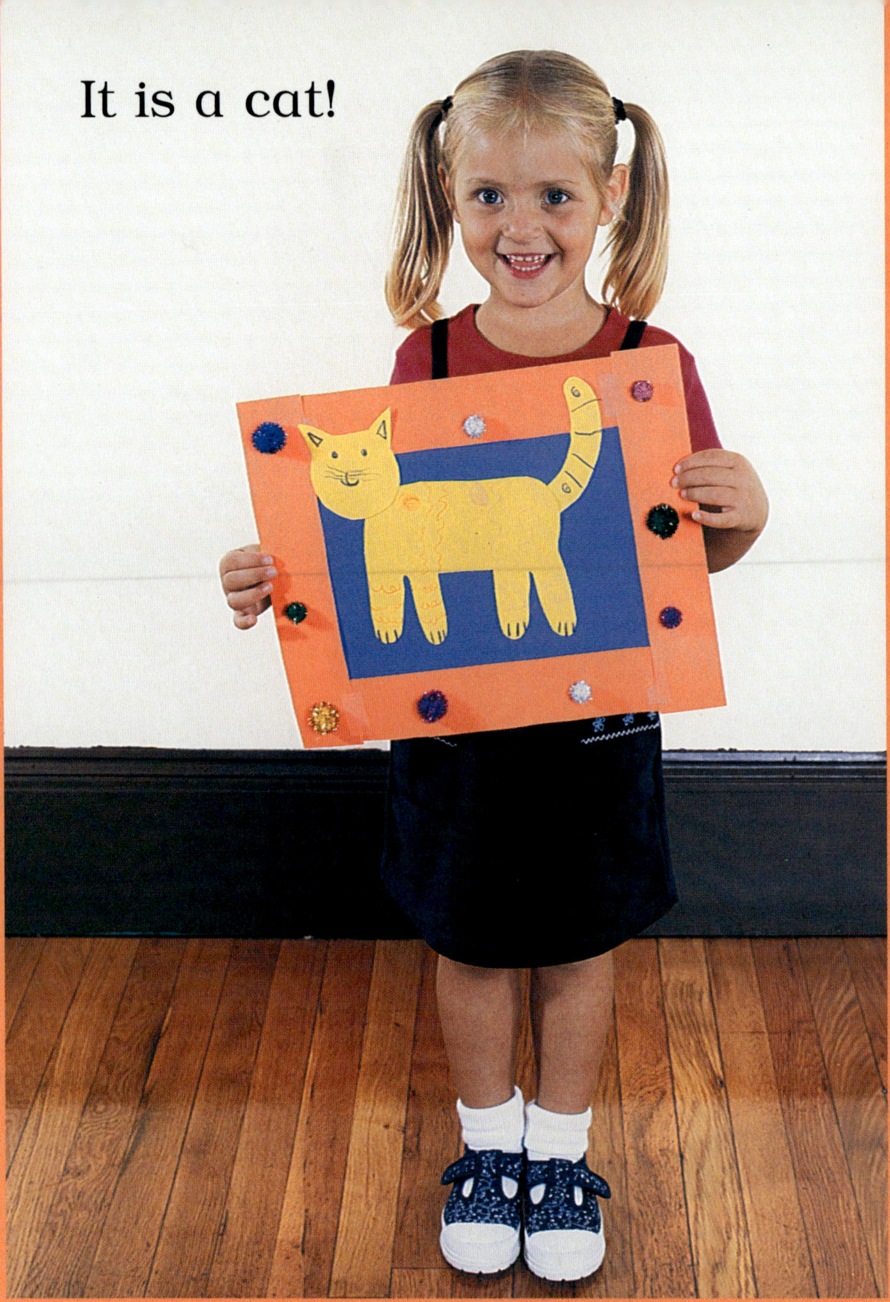